THE BIG BOOK OF GEOGRAPHY FACTS

AN EDUCATIONAL EXPLORATION PICTURE BOOK FOR KIDS ABOUT CONTINENTS, COUNTRIES, LANDFORMS, AND MANY MORE

Copyright ©2024 James K. Mahi

All rights reserved

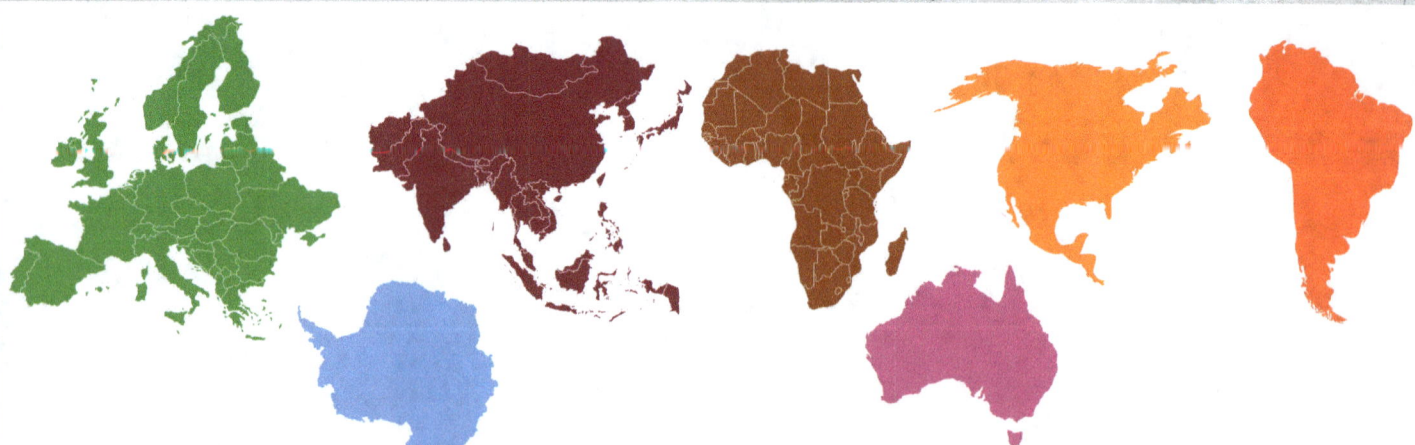

The land on Earth is divided into seven continents: Asia, Africa, North America, South America, Antarctica, Europe, and Australia.

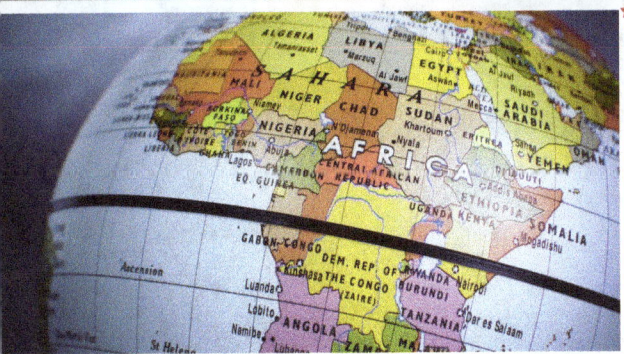

What is the name of the imaginary line that divides the Earth into the Northern and Southern Hemispheres?
The Equator.

What do we call a piece of land that is completely surrounded by water?
An island.

What country has more pyramids than Egypt?
Sudan has more pyramids than Egypt.

Which continent is home to the most countries?
Africa is home to the most countries (54).

Earth Spins Like a Top: The Earth spins around once every 24 hours, which is why we have day and night.

Earth Looks Blue from Space: Because of its oceans and atmosphere, Earth appears mostly blue when seen from space.

Antarctica is the coldest, windiest, and driest continent on the planet.

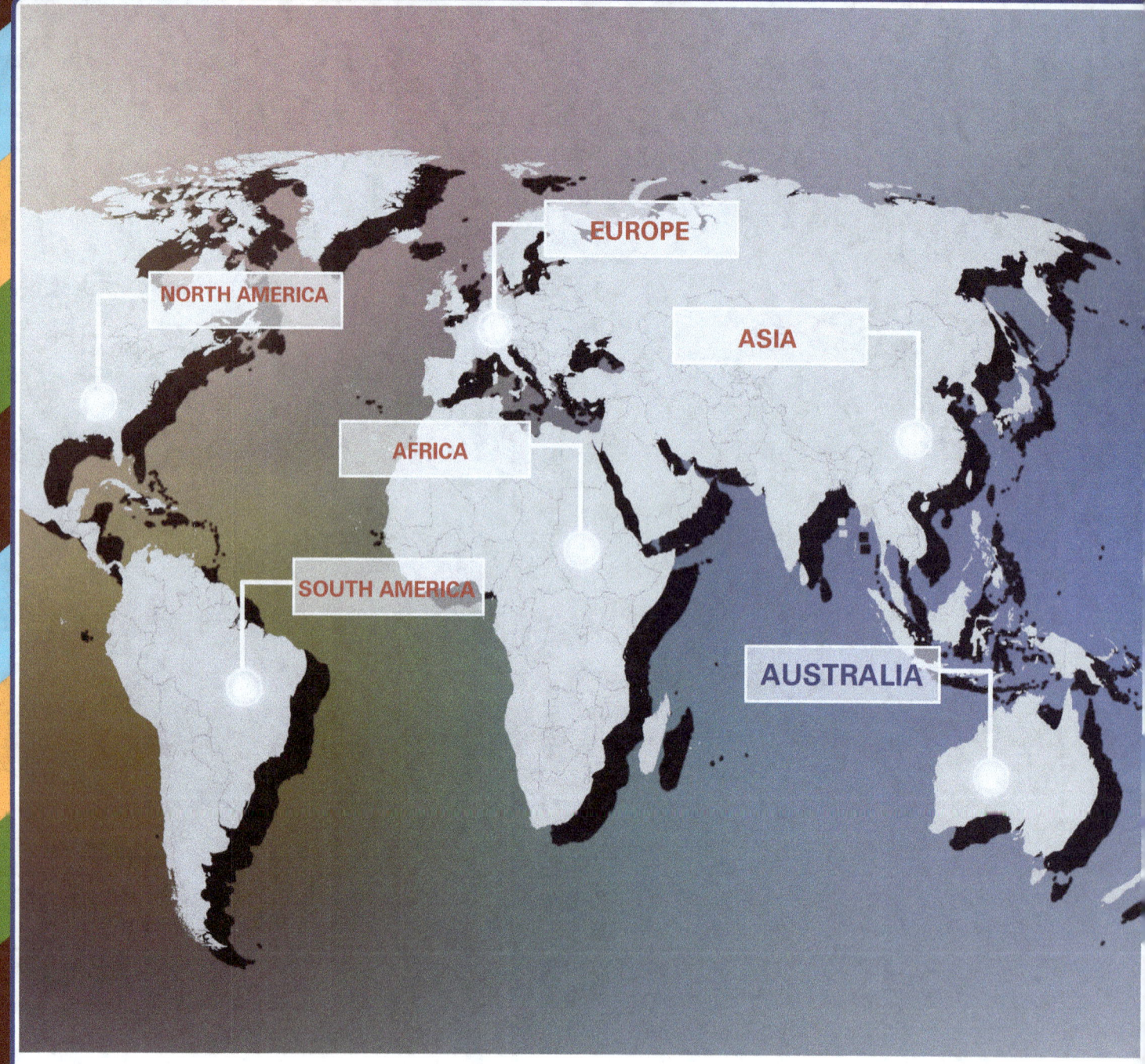

Australia is both a country and a continent.

Asia is the largest continent and has the most people living on it.

Europe is the only continent without any deserts.

The Atlantic Ocean is the second-largest ocean and has the shape of an "S."

The Southern Ocean surrounds Antarctica and is the newest ocean to be officially named (2000).

Earth Has a Magnet Field: This invisible shield protects us from harmful rays from the Sun.

North America has all climate zones: tropical, temperate, and arctic.

The Air We Breathe: Earth's atmosphere is made mostly of nitrogen (78%) and oxygen (21%), which we need to survive.

Earth is Always Moving: While it spins, Earth also travels around the Sun, taking 365 days to complete one full orbit – this gives us a year!

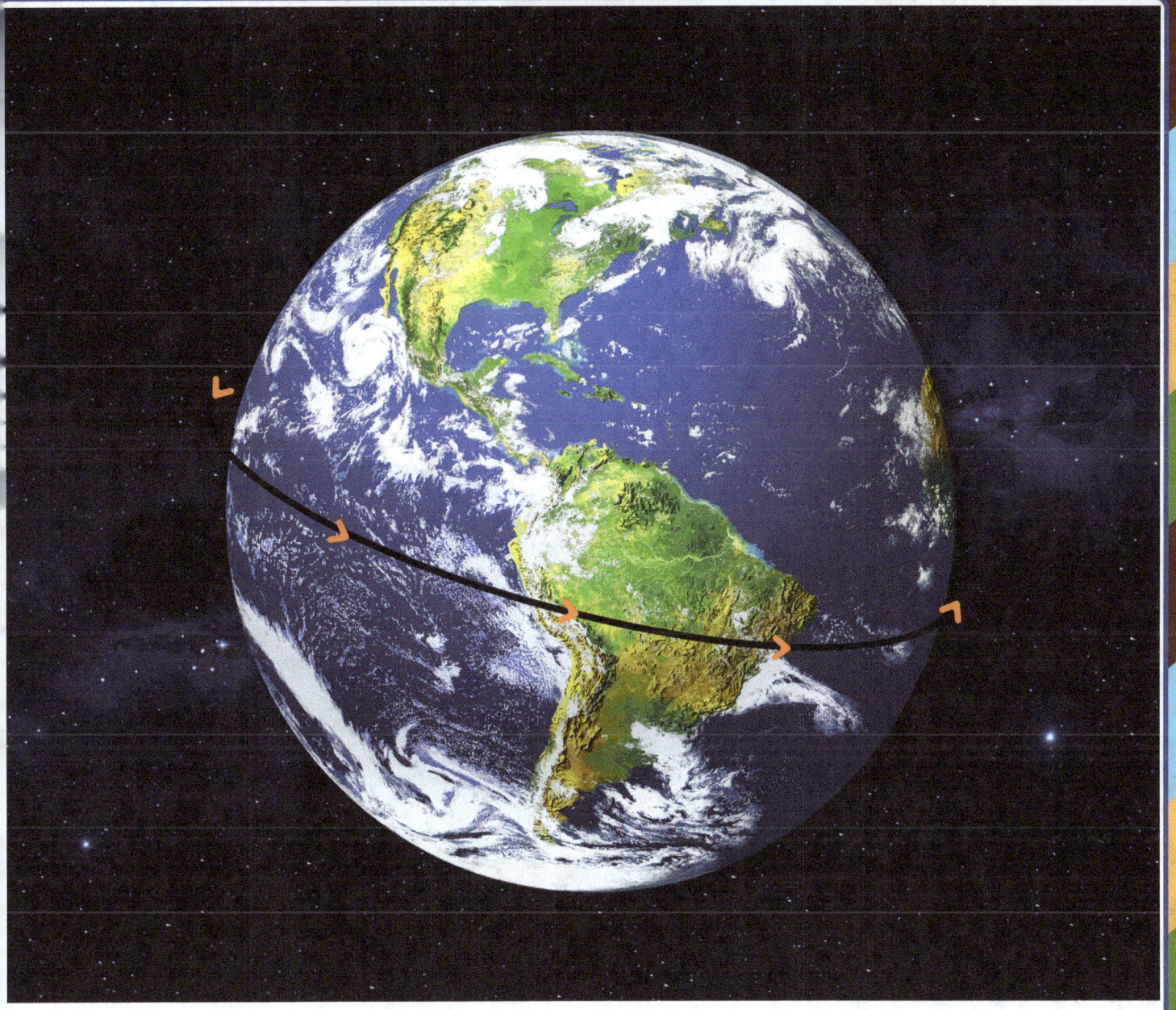

Earth Spins Like a Top: The Earth spins around once every 24 hours, which is why we have day and night.

Earth's Shape: Earth is not a perfect sphere; it's slightly flattened at the poles and bulges at the equator.

Longest River: The Nile River in Africa is about 6,650 kilometers (4,130 miles) long, making it one of the longest rivers in the world.

Antarctica is a desert because it hardly gets any rain or snow, just like a hot desert. The air there is very cold and dry, so clouds can't make much precipitation. Even though it's covered in thick ice, it almost never gets new water from rain or snow. Deserts don't have to be hot; they just have to be very dry. That's why Antarctica is the coldest and driest desert in the world!

Largest Desert: The Antarctic Desert is the largest desert in the world, even larger than the Sahara!

Deepest Ocean: The Pacific Ocean contains the Mariana Trench, with a depth of 10,994 meters (36,070 feet) at Challenger Deep.

Tallest Mountain: Mount Everest in the Himalayas stands at 8,849 meters (29,032 feet) above sea level.

Lowest Point on Land: The Dead Sea shore is the lowest land point on Earth at 430 meters (1,410 feet) below sea level.

Largest Lake: The Caspian Sea is the largest enclosed inland body of water, despite being called a sea.

Smallest Country: Vatican City is the smallest country in the world, covering only 44 hectares (110 acres).

Largest Country: Russia is the largest country, spanning 17 million square kilometers (6.6 million square miles).

Hottest Place: Death Valley in California holds the record for the hottest temperature ever recorded: 56.7°C (134°F).

Coldest Place: Antarctica has recorded temperatures as low as -89.2°C (-128.6°F).

Widest Waterfall: Khone Falls in Laos is the widest waterfall, stretching over 10.8 kilometers (6.7 miles).

Most Populous Country: India is the most populous country, with over 1.4 billion people.

Smallest Ocean: The Arctic Ocean is the smallest and shallowest of the world's oceans.

Largest Ocean : The Pacific Ocean is the largest and deepest ocean on Earth.

Highest Waterfall: Angel Falls in Venezuela is the tallest waterfall, with a height of 979 meters (3,212 feet).

Earth is 4.5 Billion Years Old: Scientists think our planet formed about 4.5 billion years ago, making it very old!

Oldest Lake: Lake Baikal is also considered the world's oldest lake, over 25 million years old.

Deepest Lake: Lake Baikal in Russia is the deepest lake, with a depth of 1,642 meters (5,387 feet).

Longest Mountain Range: The Andes in South America stretch over 7,000 kilometers (4,350 miles).

Rainforests are Earth's Lungs: The Amazon Rainforest produces about 20% of the oxygen we breathe!

Largest Volcano: Mauna Loa in Hawaii is the largest volcano on Earth.

Largest Rainforest: The Amazon Rainforest covers 5.5 million square kilometers (2.1 million square miles).

Tectonic Plates: The Earth's crust is made up of giant pieces called tectonic plates.

City on Two Continents: Istanbul is the only city located on two continents, Europe and Asia.

Longest River in Asia: The Yangtze River in China is 6,300 kilometers (3,917 miles) long.

Earth's Water: About 97% of Earth's water is in the oceans, while only 3% is fresh water.

Longest Coastline: Canada also has the longest coastline in the world, stretching over 202,000 kilometers (125,000 miles).

Driest Place: The Atacama Desert in Chile is the driest place on Earth.

Largest Coral Reef: The Great Barrier Reef in Australia is the largest coral reef system.

Country with Most Time Zones: France has the most time zones (12), including overseas territories.

Country with No Rivers: Saudi Arabia has no permanent rivers.

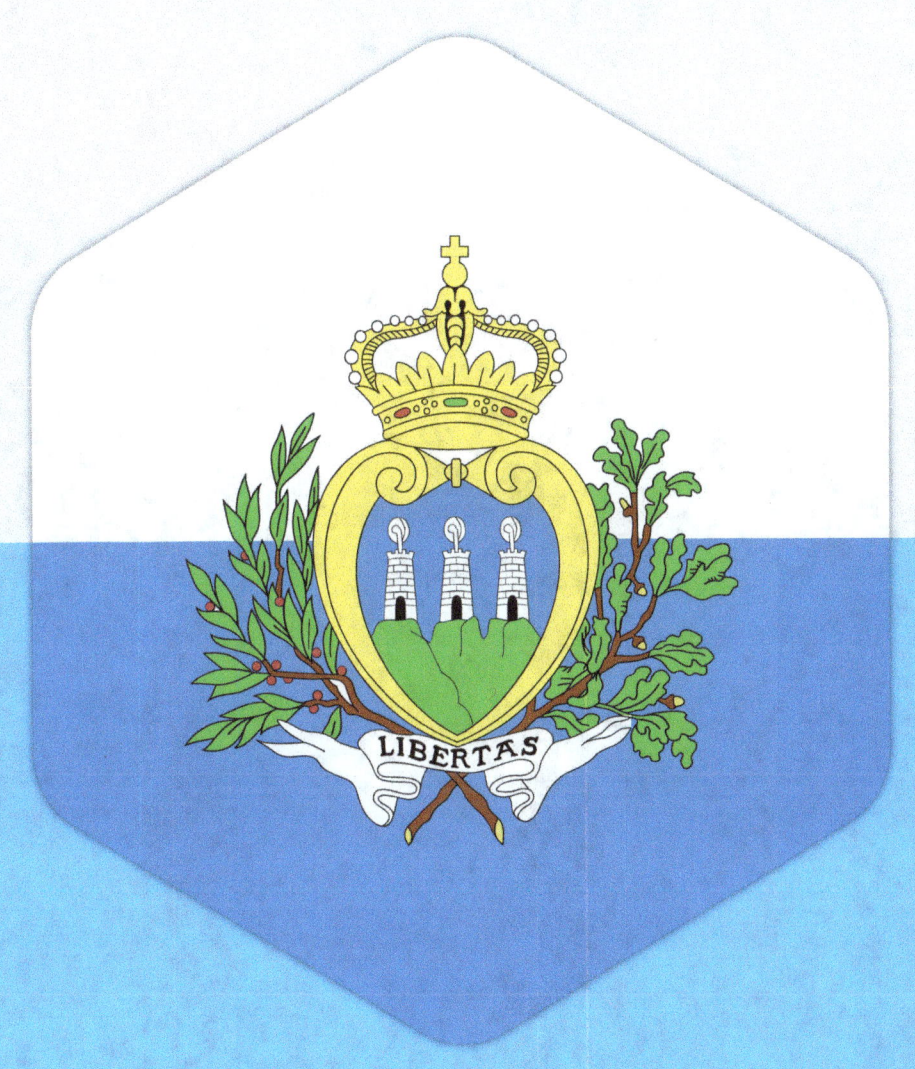

Oldest Country: San Marino claims to be the world's oldest republic, established in 301 CE.

Most Lakes: Canada has the most lakes of any country in the world.

World's Flattest Country: The Maldives has the lowest average height above sea level.

Earthquakes Shake the Ground: Earthquakes happen when tectonic plates in the Earth's crust move or crash into each other.

Earthquakes: Japan experiences about 1,500 earthquakes every year.

Tallest Trees: The tallest trees in the world, the redwoods, are found in California.

Sahara's Expansion: The Sahara Desert is expanding southward due to desertification.

Longest Border: The Canada–United States border is the longest international border.

Richest Biodiversity: Brazil is home to more plant and animal species than any other country.

Landlocked Countries: There are 44 landlocked countries in the world.

Island Continent: Australia is often called an "island continent."

Longest Highway: The Pan-American Highway is the longest, stretching from Alaska to Argentina.

Largest Cave: Son Doong Cave in Vietnam is the world's largest cave.

Floating Island: Bolivia's Salar de Uyuni becomes a giant mirror-like surface after rainfall.

Erupting Volcanoes: Indonesia has the most active volcanoes.

Largest Iceberg: The largest iceberg ever recorded was bigger than Jamaica.

Islands in the Philippines: The Philippines consists of over 7,600 islands.

Salt Flats: The Bonneville Salt Flats in Utah are so flat they are used for speed tests.

The World's Largest Island: Greenland is the biggest island in the world, even though it's covered in ice and not green!

Antarctica's Ice: Antarctica holds about 60% of the world's fresh water as ice!

Mount Vesuvius in Italy is one of the most dangerous volcanoes because of its location near Naples.

Floating Islands: The Uros people in Peru live on floating islands made of reeds on Lake Titicaca.

Earth's Busiest Ocean: The Atlantic Ocean has the most shipping traffic in the world.

Iceland has over 30 active volcanoes and sits on two tectonic plates.

Volcanic Islands: Hawaii was formed entirely by volcanic activity.

Fastest Moving Glacier: Jakobshavn Glacier in Greenland moves about 30 meters (98 feet) per day.

7 WONDERS OF THE WORLD

Great Wall of China (China)
The Great Wall is over 13,000 miles (21,000 kilometers) long and was built to protect against invasions.

Petra (Jordan)
Petra is also called the "Rose City" because of the pink-colored stone. It was carved into rock around 300 BCE.

Christ the Redeemer (Brazil)
Fact: The statue is 98 feet (30 meters) tall and stands on a mountain, offering a view of Rio de Janeiro.

Machu Picchu (Peru)
This ancient city of the Inca Empire is located 7,970 feet (2,430 meters) above sea level. It's known as the "Lost City of the Incas."

Chichen Itza (Mexico)
The Pyramid of Kukulcán, at the center of Chichen Itza, is designed so that on the equinox, a shadow forms the shape of a serpent.

Colosseum (Italy)
The Colosseum is the largest amphitheater ever built and could hold around 50,000 spectators for gladiator games.

Taj Mahal (India)

The Taj Mahal was built by Emperor Shah Jahan as a tomb for his wife Mumtaz Mahal. It's made of white marble and changes color depending on sunlight.

Thank you for exploring the wonders of our world! Remember, the Earth is full of surprises, and there's always more to learn about its amazing geography. Keep discovering!

www.ingramcontent.com/pod-product-compliance
Lightning Source LLC
Chambersburg PA
CBHW062228220526
45471CB00009B/3385